CW01584148

Dates: ______________ to ______________

Name: ____

Facility: ______________________________________

Location: ____________________________

The Observational Astronomy Skywatcher Notebook

Record 50 detailed observations of the night sky

Chris McMullen

The Observational Astronomy Skywatcher Notebook

www.faculty.lsmsa.edu/CMcMullen

Custom Books

Nonfiction / science / astronomy

Textbooks / science / astronomy

ISBN: 1438287062

EAN-13: 9781438287065

Using this notebook

Document your skywatching experiences in this astronomy notebook. Record your frame of reference, sketch key features in the night sky, record the astronomical coordinates for your observation, describe and sketch the object or event observed, describe the object or event viewed in detail, record quantitative data, and perform calculations in your notebook.

Frame of Reference: Describe your position, time, and visibility.

Sky Map: Draw and label a sketch of key celestial objects in the sky to serve as reference points for the location of the object or event you observe.

Lunar Phase: Indicate if the moon is visible in the sky. If so, draw and label the lunar phase in your sketch of the sky.

Astronomical Coordinates: The declination and right ascension specify the direction you are looking toward in the night sky from your frame of reference. The Universal Time is an important time frame to use for comparing astronomical observations with others.

Object/Event: Describe which object or event you are viewing. Draw and label a sketch of the object or event.

Observations: Describe the object or event in detail. Record quantitative measurements for the object or event such as apparent visual magnitude and size. Include units and appropriate use of significant figures.

Computations: Use this space for calculations.

Frame of Reference

Day: ________ Date: ____/____/____ Time: ________ to ________ Time Zone: __________________

Location: ____ Latitude: ________ Longitude: ________

Weather/Visibility: ____

Sky Map

Sketch and label chief celestial objects present in the sky such as planets, the ecliptic, bright stars, and constellations. Identify the region of the sky that you are viewing.

Lunar Phase

Lunar Phase: ______________________________

Visible in Sky? _____ If so, include the lunar phase in your diagram below.

Astronomical Coordinates

Record the astronomical coordinates for the region of the night sky that you are viewing.

Declination: ________________ Right Ascension: ________________ Universal Time: ________________

Object / Event

Describe which celestial object and/or event you are observing: ___________________________________

____

Sketch and label the object/event.

Observations

Describe the object/event: ____

____

____

____

____

____

____

____

____

____

____

Record quantitative data, such as apparent visual magnitude and size:

Computations

Perform calculations below.

Frame of Reference

Day: ________ Date: ____/____/____ Time: ________ to ________ Time Zone: _________________

Location: ____ Latitude: ________ Longitude: ________

Weather/Visibility: ____

Sky Map

Sketch and label chief celestial objects present in the sky such as planets, the ecliptic, bright stars, and constellations. Identify the region of the sky that you are viewing.

Lunar Phase

Lunar Phase: ______________________________

Visible in Sky? _____ If so, include the lunar phase in your diagram below.

Astronomical Coordinates

Record the astronomical coordinates for the region of the night sky that you are viewing.

Declination: ________________ Right Ascension: ________________ Universal Time: ________________

Object / Event

Describe which celestial object and/or event you are observing: ___________________________________

____

Sketch and label the object/event.

Observations

Describe the object/event: ______________________________

____

____

____

____

____

____

____

____

____

____

Record quantitative data, such as apparent visual magnitude and size:

Computations

Perform calculations below.

Frame of Reference

Day: ________ Date: ____/____/____ Time: ________ to ________ Time Zone: __________________

Location: ____ Latitude: ________ Longitude: ________

Weather/Visibility: ____

Sky Map

Sketch and label chief celestial objects present in the sky such as planets, the ecliptic, bright stars, and constellations. Identify the region of the sky that you are viewing.

Lunar Phase

Lunar Phase: ______________________________

Visible in Sky? _____ If so, include the lunar phase in your diagram below.

Astronomical Coordinates

Record the astronomical coordinates for the region of the night sky that you are viewing.

Declination: ________________ Right Ascension: ________________ Universal Time: ________________

Object / Event

Describe which celestial object and/or event you are observing: ___________________________________

____

Sketch and label the object/event.

Observations

Describe the object/event: ____

____

____

____

____

____

____

____

____

____

____

Record quantitative data, such as apparent visual magnitude and size:

Computations

Perform calculations below.

Frame of Reference

Day: ________ Date: ____/____/____ Time: ________ to ________ Time Zone: __________________

Location: ____ Latitude: ________ Longitude: ________

Weather/Visibility: ____

Sky Map

Sketch and label chief celestial objects present in the sky such as planets, the ecliptic, bright stars, and constellations. Identify the region of the sky that you are viewing.

Lunar Phase

Lunar Phase: ______________________________

Visible in Sky? _____ If so, include the lunar phase in your diagram below.

Astronomical Coordinates

Record the astronomical coordinates for the region of the night sky that you are viewing.

Declination: _______________ Right Ascension: _______________ Universal Time: _______________

Object / Event

Describe which celestial object and/or event you are observing: __________________________________

____

Sketch and label the object/event.

Observations

Describe the object/event: ____

____

____

____

____

____

____

____

____

____

____

Record quantitative data, such as apparent visual magnitude and size:

Computations

Perform calculations below.

Frame of Reference

Day: ________ Date: ____/____/____ Time: ________ to ________ Time Zone: ________________

Location: ____ Latitude: ________ Longitude: ________

Weather/Visibility: ____

Sky Map

Sketch and label chief celestial objects present in the sky such as planets, the ecliptic, bright stars, and constellations. Identify the region of the sky that you are viewing.

Lunar Phase

Lunar Phase: ____________________________

Visible in Sky? _____ If so, include the lunar phase in your diagram below.

Astronomical Coordinates

Record the astronomical coordinates for the region of the night sky that you are viewing.

Declination: _______________ Right Ascension: _______________ Universal Time: _______________

Object / Event

Describe which celestial object and/or event you are observing: ________________________________

____

Sketch and label the object/event.

Observations

Describe the object/event: ______________________________

____

____

____

____

____

____

____

____

____

____

Record quantitative data, such as apparent visual magnitude and size:

Computations

Perform calculations below.

Frame of Reference

Day: ________ Date: ____/____/____ Time: ________ to ________ Time Zone: __________________

Location: ____ Latitude: ________ Longitude: ________

Weather/Visibility: ____

Sky Map

Sketch and label chief celestial objects present in the sky such as planets, the ecliptic, bright stars, and constellations. Identify the region of the sky that you are viewing.

Lunar Phase

Lunar Phase: ______________________________

Visible in Sky? _____ If so, include the lunar phase in your diagram below.

Astronomical Coordinates

Record the astronomical coordinates for the region of the night sky that you are viewing.

Declination: ________________ Right Ascension: ________________ Universal Time: ________________

Object / Event

Describe which celestial object and/or event you are observing: ___________________________________

____

Sketch and label the object/event.

Observations

Describe the object/event: ____

____

____

____

____

____

____

____

____

____

____

Record quantitative data, such as apparent visual magnitude and size:

Computations

Perform calculations below.

Frame of Reference

Day: ________ Date: ____/____/____ Time: ________ to ________ Time Zone: __________________

Location: ____ Latitude: ________ Longitude: ________

Weather/Visibility: ____

Sky Map

Sketch and label chief celestial objects present in the sky such as planets, the ecliptic, bright stars, and constellations. Identify the region of the sky that you are viewing.

Lunar Phase

Lunar Phase: ______________________________

Visible in Sky? _____ If so, include the lunar phase in your diagram below.

Astronomical Coordinates

Record the astronomical coordinates for the region of the night sky that you are viewing.

Declination: _______________ Right Ascension: _______________ Universal Time: _______________

Object / Event

Describe which celestial object and/or event you are observing: _________________________________

____

Sketch and label the object/event.

Observations

Describe the object/event: ______________________________

Record quantitative data, such as apparent visual magnitude and size:

Computations

Perform calculations below.

Frame of Reference

Day: ________ Date: ____/____/____ Time: ________ to ________ Time Zone: __________________

Location: ____ Latitude: ________ Longitude: ________

Weather/Visibility: ____

Sky Map

Sketch and label chief celestial objects present in the sky such as planets, the ecliptic, bright stars, and constellations. Identify the region of the sky that you are viewing.

Lunar Phase

Lunar Phase: _____________________________

Visible in Sky? _____ If so, include the lunar phase in your diagram below.

Astronomical Coordinates

Record the astronomical coordinates for the region of the night sky that you are viewing.

Declination: _______________ Right Ascension: _______________ Universal Time: _______________

Object / Event

Describe which celestial object and/or event you are observing: _________________________________

____

Sketch and label the object/event.

Observations

Describe the object/event: ______________________________

____

____

____

____

____

____

____

____

____

____

Record quantitative data, such as apparent visual magnitude and size:

Computations

Perform calculations below.

Frame of Reference

Day: ________ Date: ____/____/____ Time: ________ to ________ Time Zone: __________________

Location: ____ Latitude: ________ Longitude: ________

Weather/Visibility: ____

Sky Map

Sketch and label chief celestial objects present in the sky such as planets, the ecliptic, bright stars, and constellations. Identify the region of the sky that you are viewing.

Lunar Phase

Lunar Phase: ______________________________

Visible in Sky? _____ If so, include the lunar phase in your diagram below.

Astronomical Coordinates

Record the astronomical coordinates for the region of the night sky that you are viewing.

Declination: ________________ Right Ascension: ________________ Universal Time: ________________

Object / Event

Describe which celestial object and/or event you are observing: __________________________________

____

Sketch and label the object/event.

Observations

Describe the object/event: ____

____

____

____

____

____

____

____

____

____

____

Record quantitative data, such as apparent visual magnitude and size:

Computations

Perform calculations below.

Frame of Reference

Day: ________ Date: ____/____/____ Time: ________ to ________ Time Zone: __________________

Location: ____ Latitude: ________ Longitude: ________

Weather/Visibility: ____

Sky Map

Sketch and label chief celestial objects present in the sky such as planets, the ecliptic, bright stars, and constellations. Identify the region of the sky that you are viewing.

Lunar Phase

Lunar Phase: _______________________________

Visible in Sky? _____ If so, include the lunar phase in your diagram below.

Astronomical Coordinates

Record the astronomical coordinates for the region of the night sky that you are viewing.

Declination: ________________ Right Ascension: ________________ Universal Time: ________________

Object / Event

Describe which celestial object and/or event you are observing: ___________________________________

____

Sketch and label the object/event.

Observations

Describe the object/event: ____

____

____

____

____

____

____

____

____

____

____

Record quantitative data, such as apparent visual magnitude and size:

Computations

Perform calculations below.

Frame of Reference

Day: ________ Date: ____/____/____ Time: ________ to ________ Time Zone: __________________

Location: ____ Latitude: ________ Longitude: ________

Weather/Visibility: ____

Sky Map

Sketch and label chief celestial objects present in the sky such as planets, the ecliptic, bright stars, and constellations. Identify the region of the sky that you are viewing.

Lunar Phase

Lunar Phase: ________________________________

Visible in Sky? _____ If so, include the lunar phase in your diagram below.

Astronomical Coordinates

Record the astronomical coordinates for the region of the night sky that you are viewing.

Declination: ________________ Right Ascension: ________________ Universal Time: ________________

Object / Event

Describe which celestial object and/or event you are observing: ____________________________________

____

Sketch and label the object/event.

Observations

Describe the object/event: ____

____

____

____

____

____

____

____

____

____

____

Record quantitative data, such as apparent visual magnitude and size:

Computations

Perform calculations below.

Frame of Reference

Day: ________ Date: ____/____/____ Time: ________ to ________ Time Zone: ________________

Location: ____ Latitude: ________ Longitude: ________

Weather/Visibility: ____

Sky Map

Sketch and label chief celestial objects present in the sky such as planets, the ecliptic, bright stars, and constellations. Identify the region of the sky that you are viewing.

Lunar Phase

Lunar Phase: ____________________________

Visible in Sky? _____ If so, include the lunar phase in your diagram below.

Astronomical Coordinates

Record the astronomical coordinates for the region of the night sky that you are viewing.

Declination: _______________ Right Ascension: _______________ Universal Time: _______________

Object / Event

Describe which celestial object and/or event you are observing: ________________________________

____

Sketch and label the object/event.

Observations

Describe the object/event: ____

____

____

____

____

____

____

____

____

____

____

Record quantitative data, such as apparent visual magnitude and size:

Computations

Perform calculations below.

Frame of Reference

Day: ________ Date: ____/____/____ Time: ________ to ________ Time Zone: __________________

Location: ____ Latitude: ________ Longitude: ________

Weather/Visibility: ____

Sky Map

Sketch and label chief celestial objects present in the sky such as planets, the ecliptic, bright stars, and constellations. Identify the region of the sky that you are viewing.

Lunar Phase

Lunar Phase: ______________________________

Visible in Sky? _____ If so, include the lunar phase in your diagram below.

Astronomical Coordinates

Record the astronomical coordinates for the region of the night sky that you are viewing.

Declination: ________________ Right Ascension: ________________ Universal Time: ________________

Object / Event

Describe which celestial object and/or event you are observing: __________________________________

____

Sketch and label the object/event.

Observations

Describe the object/event: ______________________________

____

____

____

____

____

____

____

____

____

____

Record quantitative data, such as apparent visual magnitude and size:

Computations

Perform calculations below.

Frame of Reference

Day: ________ Date: ____/____/____ Time: ________ to ________ Time Zone: ________________

Location: ______________________________________ Latitude: ________ Longitude: ________

Weather/Visibility: ____

Sky Map

Sketch and label chief celestial objects present in the sky such as planets, the ecliptic, bright stars, and constellations. Identify the region of the sky that you are viewing.

Lunar Phase

Lunar Phase: ____________________________

Visible in Sky? _____ If so, include the lunar phase in your diagram below.

Astronomical Coordinates

Record the astronomical coordinates for the region of the night sky that you are viewing.

Declination: _______________ Right Ascension: _______________ Universal Time: _______________

Object / Event

Describe which celestial object and/or event you are observing: ________________________________

____

Sketch and label the object/event.

Observations

Describe the object/event: ____

____

____

____

____

____

____

____

____

____

____

Record quantitative data, such as apparent visual magnitude and size:

Computations

Perform calculations below.

Frame of Reference

Day: ________ Date: ____/____/____ Time: ________ to ________ Time Zone: ________________

Location: ____ Latitude: ________ Longitude: ________

Weather/Visibility: ____

Sky Map

Sketch and label chief celestial objects present in the sky such as planets, the ecliptic, bright stars, and constellations. Identify the region of the sky that you are viewing.

Lunar Phase

Lunar Phase: ____________________________

Visible in Sky? _____ If so, include the lunar phase in your diagram below.

Astronomical Coordinates

Record the astronomical coordinates for the region of the night sky that you are viewing.

Declination: _______________ Right Ascension: _______________ Universal Time: _______________

Object / Event

Describe which celestial object and/or event you are observing: ________________________________

____

Sketch and label the object/event.

Observations

Describe the object/event: ______________________________

____

____

____

____

____

____

____

____

____

____

Record quantitative data, such as apparent visual magnitude and size:

Computations

Perform calculations below.

Frame of Reference

Day: ________ Date: ____/____/____ Time: ________ to ________ Time Zone: __________________

Location: ____ Latitude: ________ Longitude: ________

Weather/Visibility: ____

Sky Map

Sketch and label chief celestial objects present in the sky such as planets, the ecliptic, bright stars, and constellations. Identify the region of the sky that you are viewing.

Lunar Phase

Lunar Phase: _______________________________

Visible in Sky? _____ If so, include the lunar phase in your diagram below.

Astronomical Coordinates

Record the astronomical coordinates for the region of the night sky that you are viewing.

Declination: ________________ Right Ascension: ________________ Universal Time: ________________

Object / Event

Describe which celestial object and/or event you are observing: ___________________________________

____

Sketch and label the object/event.

Observations

Describe the object/event: ____

Record quantitative data, such as apparent visual magnitude and size:

Computations

Perform calculations below.

Frame of Reference

Day: ________ Date: ____/____/____ Time: ________ to ________ Time Zone: __________________

Location: ____ Latitude: ________ Longitude: ________

Weather/Visibility: ____

Sky Map

Sketch and label chief celestial objects present in the sky such as planets, the ecliptic, bright stars, and constellations. Identify the region of the sky that you are viewing.

Lunar Phase

Lunar Phase: ______________________________

Visible in Sky? _____ If so, include the lunar phase in your diagram below.

Astronomical Coordinates

Record the astronomical coordinates for the region of the night sky that you are viewing.

Declination: ________________ Right Ascension: ________________ Universal Time: ________________

Object / Event

Describe which celestial object and/or event you are observing: __________________________________

____

Sketch and label the object/event.

Observations

Describe the object/event: ______________________________

Record quantitative data, such as apparent visual magnitude and size:

Computations

Perform calculations below.

Frame of Reference

Day: ________ Date: ____/____/____ Time: ________ to ________ Time Zone: __________________

Location: ____ Latitude: ________ Longitude: ________

Weather/Visibility: ____

Sky Map

Sketch and label chief celestial objects present in the sky such as planets, the ecliptic, bright stars, and constellations. Identify the region of the sky that you are viewing.

Lunar Phase

Lunar Phase: _______________________________

Visible in Sky? _____ If so, include the lunar phase in your diagram below.

Astronomical Coordinates

Record the astronomical coordinates for the region of the night sky that you are viewing.

Declination: ________________ Right Ascension: ________________ Universal Time: ________________

Object / Event

Describe which celestial object and/or event you are observing: ___________________________________

____

Sketch and label the object/event.

Observations

Describe the object/event: ____

____

____

____

____

____

____

____

____

____

____

Record quantitative data, such as apparent visual magnitude and size:

Computations

Perform calculations below.

Frame of Reference

Day: ________ Date: ____/____/____ Time: ________ to ________ Time Zone: __________________

Location: ____ Latitude: ________ Longitude: ________

Weather/Visibility: ____

Sky Map

Sketch and label chief celestial objects present in the sky such as planets, the ecliptic, bright stars, and constellations. Identify the region of the sky that you are viewing.

Lunar Phase

Lunar Phase: ______________________________

Visible in Sky? _____ If so, include the lunar phase in your diagram below.

Astronomical Coordinates

Record the astronomical coordinates for the region of the night sky that you are viewing.

Declination: ________________ Right Ascension: ________________ Universal Time: ________________

Object / Event

Describe which celestial object and/or event you are observing: __________________________________

____

Sketch and label the object/event.

Observations

Describe the object/event: ____

____

____

____

____

____

____

____

____

____

____

Record quantitative data, such as apparent visual magnitude and size:

Computations

Perform calculations below.

Frame of Reference

Day: ________ Date: ____/____/____ Time: ________ to ________ Time Zone: ________________

Location: ____ Latitude: ________ Longitude: ________

Weather/Visibility: ____

Sky Map

Sketch and label chief celestial objects present in the sky such as planets, the ecliptic, bright stars, and constellations. Identify the region of the sky that you are viewing.

Lunar Phase

Lunar Phase: ______________________________

Visible in Sky? _____ If so, include the lunar phase in your diagram below.

Astronomical Coordinates

Record the astronomical coordinates for the region of the night sky that you are viewing.

Declination: _______________ Right Ascension: _______________ Universal Time: _______________

Object / Event

Describe which celestial object and/or event you are observing: _________________________________

____

Sketch and label the object/event.

Observations

Describe the object/event: ____

Record quantitative data, such as apparent visual magnitude and size:

Computations

Perform calculations below.

Frame of Reference

Day: ________ Date: ____/____/____ Time: ________ to ________ Time Zone: __________________

Location: ____ Latitude: ________ Longitude: ________

Weather/Visibility: ____

Sky Map

Sketch and label chief celestial objects present in the sky such as planets, the ecliptic, bright stars, and constellations. Identify the region of the sky that you are viewing.

Lunar Phase

Lunar Phase: _______________________________

Visible in Sky? _____ If so, include the lunar phase in your diagram below.

Astronomical Coordinates

Record the astronomical coordinates for the region of the night sky that you are viewing.

Declination: ________________ Right Ascension: ________________ Universal Time: ________________

Object / Event

Describe which celestial object and/or event you are observing: ___________________________________

____

Sketch and label the object/event.

Observations

Describe the object/event: ______________________________

____

____

____

____

____

____

____

____

____

____

Record quantitative data, such as apparent visual magnitude and size:

Computations

Perform calculations below.

Frame of Reference

Day: ________ Date: ____/____/____ Time: ________ to ________ Time Zone: __________________

Location: ____ Latitude: ________ Longitude: ________

Weather/Visibility: ____

Sky Map

Sketch and label chief celestial objects present in the sky such as planets, the ecliptic, bright stars, and constellations. Identify the region of the sky that you are viewing.

Lunar Phase

Lunar Phase: ______________________________

Visible in Sky? _____ If so, include the lunar phase in your diagram below.

Astronomical Coordinates

Record the astronomical coordinates for the region of the night sky that you are viewing.

Declination: ________________ Right Ascension: ________________ Universal Time: ________________

Object / Event

Describe which celestial object and/or event you are observing: __________________________________

____

Sketch and label the object/event.

Observations

Describe the object/event: ______________________________

____

____

____

____

____

____

____

____

____

____

Record quantitative data, such as apparent visual magnitude and size:

Computations

Perform calculations below.

Frame of Reference

Day: ________ Date: ____/____/____ Time: ________ to ________ Time Zone: __________________

Location: ____ Latitude: ________ Longitude: ________

Weather/Visibility: ____

Sky Map

Sketch and label chief celestial objects present in the sky such as planets, the ecliptic, bright stars, and constellations. Identify the region of the sky that you are viewing.

Lunar Phase

Lunar Phase: ______________________________

Visible in Sky? _____ If so, include the lunar phase in your diagram below.

Astronomical Coordinates

Record the astronomical coordinates for the region of the night sky that you are viewing.

Declination: ________________ Right Ascension: ________________ Universal Time: ________________

Object / Event

Describe which celestial object and/or event you are observing: __________________________________

____

Sketch and label the object/event.

Observations

Describe the object/event: ____

____

____

____

____

____

____

____

____

____

____

Record quantitative data, such as apparent visual magnitude and size:

Computations

Perform calculations below.

Frame of Reference

Day: ________ Date: ____/____/____ Time: ________ to ________ Time Zone: __________________

Location: ____ Latitude: ________ Longitude: ________

Weather/Visibility: ____

Sky Map

Sketch and label chief celestial objects present in the sky such as planets, the ecliptic, bright stars, and constellations. Identify the region of the sky that you are viewing.

Lunar Phase

Lunar Phase: ______________________________

Visible in Sky? _____ If so, include the lunar phase in your diagram below.

Astronomical Coordinates

Record the astronomical coordinates for the region of the night sky that you are viewing.

Declination: ________________ Right Ascension: ________________ Universal Time: ________________

Object / Event

Describe which celestial object and/or event you are observing: ___________________________________

____

Sketch and label the object/event.

Observations

Describe the object/event: ______________________________

____

____

____

____

____

____

____

____

____

____

Record quantitative data, such as apparent visual magnitude and size:

Computations

Perform calculations below.

Frame of Reference

Day: ________ Date: ____/____/____ Time: ________ to ________ Time Zone: ________________

Location: ____ Latitude: ________ Longitude: ________

Weather/Visibility: ____

Sky Map

Sketch and label chief celestial objects present in the sky such as planets, the ecliptic, bright stars, and constellations. Identify the region of the sky that you are viewing.

Lunar Phase

Lunar Phase: ______________________________

Visible in Sky? _____ If so, include the lunar phase in your diagram below.

Astronomical Coordinates

Record the astronomical coordinates for the region of the night sky that you are viewing.

Declination: ________________ Right Ascension: ________________ Universal Time: ________________

Object / Event

Describe which celestial object and/or event you are observing: ___________________________________

____

Sketch and label the object/event.

Observations

Describe the object/event: ____________________

____

____

____

____

____

____

____

____

____

____

Record quantitative data, such as apparent visual magnitude and size:

Computations

Perform calculations below.

Frame of Reference

Day: ________ Date: ____/____/____ Time: ________ to ________ Time Zone: __________________

Location: ____ Latitude: ________ Longitude: ________

Weather/Visibility: ____

Sky Map

Sketch and label chief celestial objects present in the sky such as planets, the ecliptic, bright stars, and constellations. Identify the region of the sky that you are viewing.

Lunar Phase

Lunar Phase: ______________________________

Visible in Sky? _____ If so, include the lunar phase in your diagram below.

Astronomical Coordinates

Record the astronomical coordinates for the region of the night sky that you are viewing.

Declination: ________________ Right Ascension: ________________ Universal Time: ________________

Object / Event

Describe which celestial object and/or event you are observing: ___________________________________

____

Sketch and label the object/event.

Observations

Describe the object/event: ______________________________

Record quantitative data, such as apparent visual magnitude and size:

Computations

Perform calculations below.

Frame of Reference

Day: ________ Date: ____/____/____ Time: ________ to ________ Time Zone: ________________

Location: ____ Latitude: ________ Longitude: ________

Weather/Visibility: ____

Sky Map

Sketch and label chief celestial objects present in the sky such as planets, the ecliptic, bright stars, and constellations. Identify the region of the sky that you are viewing.

Lunar Phase

Lunar Phase: ______________________________

Visible in Sky? _____ If so, include the lunar phase in your diagram below.

Astronomical Coordinates

Record the astronomical coordinates for the region of the night sky that you are viewing.

Declination: ________________ Right Ascension: ________________ Universal Time: ________________

Object / Event

Describe which celestial object and/or event you are observing: ___________________________________

____

Sketch and label the object/event.

Observations

Describe the object/event: ____

Record quantitative data, such as apparent visual magnitude and size:

Computations

Perform calculations below.

Frame of Reference

Day: ________ Date: ____/____/____ Time: ________ to ________ Time Zone: __________________

Location: ____ Latitude: ________ Longitude: ________

Weather/Visibility: ____

Sky Map

Sketch and label chief celestial objects present in the sky such as planets, the ecliptic, bright stars, and constellations. Identify the region of the sky that you are viewing.

Lunar Phase

Lunar Phase: _______________________________

Visible in Sky? _____ If so, include the lunar phase in your diagram below.

Astronomical Coordinates

Record the astronomical coordinates for the region of the night sky that you are viewing.

Declination: ________________ Right Ascension: ________________ Universal Time: ________________

Object / Event

Describe which celestial object and/or event you are observing: ___________________________________

____

Sketch and label the object/event.

Observations

Describe the object/event: ____

Record quantitative data, such as apparent visual magnitude and size:

Computations

Perform calculations below.

Frame of Reference

Day: ________ Date: ____/____/____ Time: ________ to ________ Time Zone: __________________

Location: ____ Latitude: ________ Longitude: ________

Weather/Visibility: ____

Sky Map

Sketch and label chief celestial objects present in the sky such as planets, the ecliptic, bright stars, and constellations. Identify the region of the sky that you are viewing.

Lunar Phase

Lunar Phase: ______________________________

Visible in Sky? _____ If so, include the lunar phase in your diagram below.

Astronomical Coordinates

Record the astronomical coordinates for the region of the night sky that you are viewing.

Declination: ________________ Right Ascension: ________________ Universal Time: ________________

Object / Event

Describe which celestial object and/or event you are observing: __________________________________

____

Sketch and label the object/event.

Observations

Describe the object/event: ______________________________

____

____

____

____

____

____

____

____

____

____

Record quantitative data, such as apparent visual magnitude and size:

Computations

Perform calculations below.

Frame of Reference

Day: ________ Date: ____/____/____ Time: ________ to ________ Time Zone: __________________

Location: ____ Latitude: ________ Longitude: ________

Weather/Visibility: ____

Sky Map

Sketch and label chief celestial objects present in the sky such as planets, the ecliptic, bright stars, and constellations. Identify the region of the sky that you are viewing.

Lunar Phase

Lunar Phase: ______________________________

Visible in Sky? _____ If so, include the lunar phase in your diagram below.

Astronomical Coordinates

Record the astronomical coordinates for the region of the night sky that you are viewing.

Declination: ________________ Right Ascension: ________________ Universal Time: ________________

Object / Event

Describe which celestial object and/or event you are observing: __________________________________

____

Sketch and label the object/event.

Observations

Describe the object/event: ______________________________

Record quantitative data, such as apparent visual magnitude and size:

Computations

Perform calculations below.

Frame of Reference

Day: ________ Date: ____/____/____ Time: ________ to ________ Time Zone: __________________

Location: ____ Latitude: ________ Longitude: ________

Weather/Visibility: ____

Sky Map

Sketch and label chief celestial objects present in the sky such as planets, the ecliptic, bright stars, and constellations. Identify the region of the sky that you are viewing.

Lunar Phase

Lunar Phase: ________________________________

Visible in Sky? _____ If so, include the lunar phase in your diagram below.

Astronomical Coordinates

Record the astronomical coordinates for the region of the night sky that you are viewing.

Declination: ________________ Right Ascension: ________________ Universal Time: ________________

Object / Event

Describe which celestial object and/or event you are observing: ____________________________________

____

Sketch and label the object/event.

Observations

Describe the object/event: ____

____

____

____

____

____

____

____

____

____

____

Record quantitative data, such as apparent visual magnitude and size:

Computations

Perform calculations below.

Frame of Reference

Day: ________ Date: ____/____/____ Time: ________ to ________ Time Zone: __________________

Location: ____ Latitude: ________ Longitude: ________

Weather/Visibility: ____

Sky Map

Sketch and label chief celestial objects present in the sky such as planets, the ecliptic, bright stars, and constellations. Identify the region of the sky that you are viewing.

Lunar Phase

Lunar Phase: _______________________________

Visible in Sky? _____ If so, include the lunar phase in your diagram below.

Astronomical Coordinates

Record the astronomical coordinates for the region of the night sky that you are viewing.

Declination: ________________ Right Ascension: ________________ Universal Time: ________________

Object / Event

Describe which celestial object and/or event you are observing: ___________________________________

____

Sketch and label the object/event.

Observations

Describe the object/event: ____

____

____

____

____

____

____

____

____

____

____

Record quantitative data, such as apparent visual magnitude and size:

Computations

Perform calculations below.

Frame of Reference

Day: ________ Date: ____/____/____ Time: ________ to ________ Time Zone: __________________

Location: ____ Latitude: ________ Longitude: ________

Weather/Visibility: ____

Sky Map

Sketch and label chief celestial objects present in the sky such as planets, the ecliptic, bright stars, and constellations. Identify the region of the sky that you are viewing.

Lunar Phase

Lunar Phase: _______________________________

Visible in Sky? _____ If so, include the lunar phase in your diagram below.

Astronomical Coordinates

Record the astronomical coordinates for the region of the night sky that you are viewing.

Declination: ________________ Right Ascension: ________________ Universal Time: ________________

Object / Event

Describe which celestial object and/or event you are observing: ___________________________________

____

Sketch and label the object/event.

Observations

Describe the object/event: ____

____

____

____

____

____

____

____

____

____

____

Record quantitative data, such as apparent visual magnitude and size:

Computations

Perform calculations below.

Frame of Reference

Day: ________ Date: ____/____/____ Time: ________ to ________ Time Zone: __________________

Location: ____ Latitude: ________ Longitude: ________

Weather/Visibility: ____

Sky Map

Sketch and label chief celestial objects present in the sky such as planets, the ecliptic, bright stars, and constellations. Identify the region of the sky that you are viewing.

Lunar Phase

Lunar Phase: ______________________________

Visible in Sky? _____ If so, include the lunar phase in your diagram below.

Astronomical Coordinates

Record the astronomical coordinates for the region of the night sky that you are viewing.

Declination: ________________ Right Ascension: ________________ Universal Time: ________________

Object / Event

Describe which celestial object and/or event you are observing: __________________________________

____

Sketch and label the object/event.

Observations

Describe the object/event: ______________________________

Record quantitative data, such as apparent visual magnitude and size:

Computations

Perform calculations below.

Frame of Reference

Day: ________ Date: ___/___/___ Time: ________ to ________ Time Zone: ______________

Location: ________________________________ Latitude: ________ Longitude: ________

Weather/Visibility: ____

Sky Map

Sketch and label chief celestial objects present in the sky such as planets, the ecliptic, bright stars, and constellations. Identify the region of the sky that you are viewing.

Lunar Phase

Lunar Phase: ______________________

Visible in Sky? _____ If so, include the lunar phase in your diagram below.

Astronomical Coordinates

Record the astronomical coordinates for the region of the night sky that you are viewing.

Declination: ______________ Right Ascension: ______________ Universal Time: ______________

Object / Event

Describe which celestial object and/or event you are observing: ______________________

____

Sketch and label the object/event.

Observations

Describe the object/event: ______________________________

____

____

____

____

____

____

____

____

____

____

Record quantitative data, such as apparent visual magnitude and size:

Computations

Perform calculations below.

Frame of Reference

Day: ________ Date: ____/____/____ Time: ________ to ________ Time Zone: __________________

Location: ____ Latitude: ________ Longitude: ________

Weather/Visibility: ____

Sky Map

Sketch and label chief celestial objects present in the sky such as planets, the ecliptic, bright stars, and constellations. Identify the region of the sky that you are viewing.

Lunar Phase

Lunar Phase: ________________________________

Visible in Sky? _____ If so, include the lunar phase in your diagram below.

Astronomical Coordinates

Record the astronomical coordinates for the region of the night sky that you are viewing.

Declination: ________________ Right Ascension: ________________ Universal Time: ________________

Object / Event

Describe which celestial object and/or event you are observing: ___________________________________

____

Sketch and label the object/event.

Observations

Describe the object/event: ____

____

____

____

____

____

____

____

____

____

____

Record quantitative data, such as apparent visual magnitude and size:

Computations

Perform calculations below.

Frame of Reference

Day: ________ Date: ____/____/____ Time: ________ to ________ Time Zone: __________________

Location: ____ Latitude: ________ Longitude: ________

Weather/Visibility: ____

Sky Map

Sketch and label chief celestial objects present in the sky such as planets, the ecliptic, bright stars, and constellations. Identify the region of the sky that you are viewing.

Lunar Phase

Lunar Phase: _______________________________

Visible in Sky? _____ If so, include the lunar phase in your diagram below.

Astronomical Coordinates

Record the astronomical coordinates for the region of the night sky that you are viewing.

Declination: ________________ Right Ascension: ________________ Universal Time: ________________

Object / Event

Describe which celestial object and/or event you are observing: ___________________________________

____

Sketch and label the object/event.

Observations

Describe the object/event: ____

____

____

____

____

____

____

____

____

____

____

Record quantitative data, such as apparent visual magnitude and size:

Computations

Perform calculations below.

Frame of Reference

Day: ________ Date: ____/____/____ Time: ________ to ________ Time Zone: __________________

Location: ____ Latitude: ________ Longitude: ________

Weather/Visibility: ____

Sky Map

Sketch and label chief celestial objects present in the sky such as planets, the ecliptic, bright stars, and constellations. Identify the region of the sky that you are viewing.

Lunar Phase

Lunar Phase: ______________________________

Visible in Sky? _____ If so, include the lunar phase in your diagram below.

Astronomical Coordinates

Record the astronomical coordinates for the region of the night sky that you are viewing.

Declination: ________________ Right Ascension: ________________ Universal Time: ________________

Object / Event

Describe which celestial object and/or event you are observing: __________________________________

____

Sketch and label the object/event.

Observations

Describe the object/event: ____

____

____

____

____

____

____

____

____

____

____

Record quantitative data, such as apparent visual magnitude and size:

Computations

Perform calculations below.

Frame of Reference

Day: ________ Date: ____/____/____ Time: ________ to ________ Time Zone: __________________

Location: ____ Latitude: ________ Longitude: ________

Weather/Visibility: ____

Sky Map

Sketch and label chief celestial objects present in the sky such as planets, the ecliptic, bright stars, and constellations. Identify the region of the sky that you are viewing.

Lunar Phase

Lunar Phase: _______________________________

Visible in Sky? _____ If so, include the lunar phase in your diagram below.

Astronomical Coordinates

Record the astronomical coordinates for the region of the night sky that you are viewing.

Declination: ________________ Right Ascension: ________________ Universal Time: ________________

Object / Event

Describe which celestial object and/or event you are observing: ___________________________________

____

Sketch and label the object/event.

Observations

Describe the object/event: ______________________________

Record quantitative data, such as apparent visual magnitude and size:

Computations

Perform calculations below.

Frame of Reference

Day: ________ Date: ____/____/____ Time: ________ to ________ Time Zone: __________________

Location: ____ Latitude: ________ Longitude: ________

Weather/Visibility: ____

Sky Map

Sketch and label chief celestial objects present in the sky such as planets, the ecliptic, bright stars, and constellations. Identify the region of the sky that you are viewing.

Lunar Phase

Lunar Phase: ______________________________

Visible in Sky? _____ If so, include the lunar phase in your diagram below.

Astronomical Coordinates

Record the astronomical coordinates for the region of the night sky that you are viewing.

Declination: ________________ Right Ascension: ________________ Universal Time: ________________

Object / Event

Describe which celestial object and/or event you are observing: ______________________________

____

Sketch and label the object/event.

Observations

Describe the object/event: ____

____

____

____

____

____

____

____

____

____

____

Record quantitative data, such as apparent visual magnitude and size:

Computations

Perform calculations below.

Frame of Reference

Day: ________ Date: ____/____/____ Time: ________ to ________ Time Zone: ________________

Location: ____ Latitude: ________ Longitude: ________

Weather/Visibility: ____

Sky Map

Sketch and label chief celestial objects present in the sky such as planets, the ecliptic, bright stars, and constellations. Identify the region of the sky that you are viewing.

Lunar Phase

Lunar Phase: ____________________________

Visible in Sky? _____ If so, include the lunar phase in your diagram below.

Astronomical Coordinates

Record the astronomical coordinates for the region of the night sky that you are viewing.

Declination: _______________ Right Ascension: _______________ Universal Time: _______________

Object / Event

Describe which celestial object and/or event you are observing: ________________________________

____

Sketch and label the object/event.

Observations

Describe the object/event: ____

Record quantitative data, such as apparent visual magnitude and size:

Computations

Perform calculations below.

Frame of Reference

Day: ________ Date: ____/____/____ Time: ________ to ________ Time Zone: ________________

Location: ____ Latitude: ________ Longitude: ________

Weather/Visibility: ____

Sky Map

Sketch and label chief celestial objects present in the sky such as planets, the ecliptic, bright stars, and constellations. Identify the region of the sky that you are viewing.

Lunar Phase

Lunar Phase: ____________________________

Visible in Sky? _____ If so, include the lunar phase in your diagram below.

Astronomical Coordinates

Record the astronomical coordinates for the region of the night sky that you are viewing.

Declination: _______________ Right Ascension: _______________ Universal Time: _______________

Object / Event

Describe which celestial object and/or event you are observing: ________________________________

____

Sketch and label the object/event.

Observations

Describe the object/event: ______________________________

____

____

____

____

____

____

____

____

____

____

Record quantitative data, such as apparent visual magnitude and size:

Computations

Perform calculations below.

Frame of Reference

Day: ________ Date: ____/____/____ Time: ________ to ________ Time Zone: __________________

Location: ____ Latitude: ________ Longitude: ________

Weather/Visibility: ____

Sky Map

Sketch and label chief celestial objects present in the sky such as planets, the ecliptic, bright stars, and constellations. Identify the region of the sky that you are viewing.

Lunar Phase

Lunar Phase: _______________________________

Visible in Sky? _____ If so, include the lunar phase in your diagram below.

Astronomical Coordinates

Record the astronomical coordinates for the region of the night sky that you are viewing.

Declination: ________________ Right Ascension: ________________ Universal Time: ________________

Object / Event

Describe which celestial object and/or event you are observing: ___________________________________

____

Sketch and label the object/event.

Observations

Describe the object/event: ____

____

____

____

____

____

____

____

____

____

____

Record quantitative data, such as apparent visual magnitude and size:

Computations

Perform calculations below.

Frame of Reference

Day: ________ Date: ____/____/____ Time: ________ to ________ Time Zone: __________________

Location: ____ Latitude: ________ Longitude: ________

Weather/Visibility: ____

Sky Map

Sketch and label chief celestial objects present in the sky such as planets, the ecliptic, bright stars, and constellations. Identify the region of the sky that you are viewing.

Lunar Phase

Lunar Phase: ______________________________

Visible in Sky? _____ If so, include the lunar phase in your diagram below.

Astronomical Coordinates

Record the astronomical coordinates for the region of the night sky that you are viewing.

Declination: ________________ Right Ascension: ________________ Universal Time: ________________

Object / Event

Describe which celestial object and/or event you are observing: ___________________________________

____

Sketch and label the object/event.

Observations

Describe the object/event: ____

____

____

____

____

____

____

____

____

____

____

Record quantitative data, such as apparent visual magnitude and size:

Computations

Perform calculations below.

Frame of Reference

Day: ________ Date: ____/____/____ Time: ________ to ________ Time Zone: __________________

Location: ____ Latitude: ________ Longitude: ________

Weather/Visibility: ____

Sky Map

Sketch and label chief celestial objects present in the sky such as planets, the ecliptic, bright stars, and constellations. Identify the region of the sky that you are viewing.

Lunar Phase

Lunar Phase: ______________________________

Visible in Sky? _____ If so, include the lunar phase in your diagram below.

Astronomical Coordinates

Record the astronomical coordinates for the region of the night sky that you are viewing.

Declination: ________________ Right Ascension: ________________ Universal Time: ________________

Object / Event

Describe which celestial object and/or event you are observing: __________________________________

____

Sketch and label the object/event.

Observations

Describe the object/event: ____

____

____

____

____

____

____

____

____

____

____

Record quantitative data, such as apparent visual magnitude and size:

Computations

Perform calculations below.

Frame of Reference

Day: ________ Date: ____/____/____ Time: ________ to ________ Time Zone: __________________

Location: ____ Latitude: ________ Longitude: ________

Weather/Visibility: ____

Sky Map

Sketch and label chief celestial objects present in the sky such as planets, the ecliptic, bright stars, and constellations. Identify the region of the sky that you are viewing.

Lunar Phase

Lunar Phase: ______________________________

Visible in Sky? _____ If so, include the lunar phase in your diagram below.

Astronomical Coordinates

Record the astronomical coordinates for the region of the night sky that you are viewing.

Declination: ________________ Right Ascension: ________________ Universal Time: ________________

Object / Event

Describe which celestial object and/or event you are observing: __________________________________

____

Sketch and label the object/event.

Observations

Describe the object/event: ______________________________

____

____

____

____

____

____

____

____

____

____

Record quantitative data, such as apparent visual magnitude and size:

Computations

Perform calculations below.

Frame of Reference

Day: ________ Date: ____/____/____ Time: ________ to ________ Time Zone: ________________

Location: ____ Latitude: ________ Longitude: ________

Weather/Visibility: ____

Sky Map

Sketch and label chief celestial objects present in the sky such as planets, the ecliptic, bright stars, and constellations. Identify the region of the sky that you are viewing.

Lunar Phase

Lunar Phase: ______________________________

Visible in Sky? _____ If so, include the lunar phase in your diagram below.

Astronomical Coordinates

Record the astronomical coordinates for the region of the night sky that you are viewing.

Declination: ________________ Right Ascension: ________________ Universal Time: ________________

Object / Event

Describe which celestial object and/or event you are observing: __________________________________

____

Sketch and label the object/event.

Observations

Describe the object/event: ____

____

____

____

____

____

____

____

____

____

____

Record quantitative data, such as apparent visual magnitude and size:

Computations

Perform calculations below.

Frame of Reference

Day: ________ Date: ____/____/____ Time: ________ to ________ Time Zone: __________________

Location: ____ Latitude: ________ Longitude: ________

Weather/Visibility: ____

Sky Map

Sketch and label chief celestial objects present in the sky such as planets, the ecliptic, bright stars, and constellations. Identify the region of the sky that you are viewing.

Lunar Phase

Lunar Phase: _______________________________

Visible in Sky? _____ If so, include the lunar phase in your diagram below.

Astronomical Coordinates

Record the astronomical coordinates for the region of the night sky that you are viewing.

Declination: ________________ Right Ascension: ________________ Universal Time: ________________

Object / Event

Describe which celestial object and/or event you are observing: ___________________________________

____

Sketch and label the object/event.

Observations

Describe the object/event: ____

____

____

____

____

____

____

____

____

____

____

Record quantitative data, such as apparent visual magnitude and size:

Computations

Perform calculations below.

Frame of Reference

Day: ________ Date: ____/____/____ Time: ________ to ________ Time Zone: ________________

Location: ____ Latitude: ________ Longitude: ________

Weather/Visibility: ____

Sky Map

Sketch and label chief celestial objects present in the sky such as planets, the ecliptic, bright stars, and constellations. Identify the region of the sky that you are viewing.

Lunar Phase

Lunar Phase: ______________________________

Visible in Sky? _____ If so, include the lunar phase in your diagram below.

Astronomical Coordinates

Record the astronomical coordinates for the region of the night sky that you are viewing.

Declination: ________________ Right Ascension: ________________ Universal Time: ________________

Object / Event

Describe which celestial object and/or event you are observing: ______________________________

____

Sketch and label the object/event.

Observations

Describe the object/event: ____

____

____

____

____

____

____

____

____

____

____

Record quantitative data, such as apparent visual magnitude and size:

Computations

Perform calculations below.

Frame of Reference

Day: ________ Date: ____/____/____ Time: ________ to ________ Time Zone: __________________

Location: ____ Latitude: ________ Longitude: ________

Weather/Visibility: ____

Sky Map

Sketch and label chief celestial objects present in the sky such as planets, the ecliptic, bright stars, and constellations. Identify the region of the sky that you are viewing.

Lunar Phase

Lunar Phase: ______________________________

Visible in Sky? _____ If so, include the lunar phase in your diagram below.

Astronomical Coordinates

Record the astronomical coordinates for the region of the night sky that you are viewing.

Declination: _______________ Right Ascension: _______________ Universal Time: _______________

Object / Event

Describe which celestial object and/or event you are observing: _________________________________

____

Sketch and label the object/event.

Observations

Describe the object/event: ____

____

____

____

____

____

____

____

____

____

____

Record quantitative data, such as apparent visual magnitude and size:

Computations

Perform calculations below.

Notes

Notes

Notes

Notes

Notes

Notes

Notes

8822514R00062

Printed in Great Britain
by Amazon.co.uk, Ltd.,
Marston Gate.